Discovering

SHREWS, MOLES & VOLES

Jill Bailey

Illustrated by
John Yates

The Bookwright Press
New York · 1989

Discovering Nature

Discovering Ants
Discovering Bats
Discovering Bees and Wasps
Discovering Beetles
Discovering Birds of Prey
Discovering Bugs
Discovering Butterflies and Moths
Discovering Crabs and Lobsters
Discovering Crickets and Grasshoppers
Discovering Damselflies and Dragonflies
Discovering Deer
Discovering Ducks, Geese and Swans
Discovering Flies
Discovering Flowering Plants
Discovering Foxes

Discovering Freshwater Fish
Discovering Frogs and Toads
Discovering Fungi
Discovering Jellyfish
Discovering Rabbits and Hares
Discovering Rats and Mice
Discovering Saltwater Fish
Discovering Sea Birds
Discovering Shrews, Moles and Voles
Discovering Slugs and Snails
Discovering Snakes and Lizards
Discovering Spiders
Discovering Squirrels
Discovering Trees
Discovering Weasels
Discovering Worms

First published in the
United States in 1989 by
The Bookwright Press
387 Park Avenue South
New York, NY 10016

First published in 1989 by
Wayland (Publishers) Limited
61 Western Road, Hove
East Sussex BN3 1JD, England

Frontispiece *A northern water shrew sits in the sun beside a river in Colorado.*

Cover *A short-tailed shrew looking for food.*

Library of Congress Cataloging-in-Publication Data
Bailey, Jill
   Discovering shrews, moles and voles / by Jill Bailey.
     p. cm. — (Discovering nature)
   Bibliography: p.
   Includes index.
  Summary: Examines the characteristics, eating habits,
habitats, behavior, and social life of shrews, moles and voles.
   ISBN 0–531–18291–6
   1. Shrews — Juvenile literature. 2. Moles (Animals) —
Juvenile literature. 3. Voles — Juvenile literature. [1. Shrews.
2. Moles (Animals) 3. Voles.] I. Title. II. Series.
QL737.I56B35   1989
599.3'3 — dc19
                                      89–30998
                                      CIP
                                      AC

Typeset by DP Press Ltd., Sevenoaks, Kent
Printed in Italy by Sagdos S.p.A., Milan

Contents

1
Introducing Shrews, Moles and Voles

The pygmy shrew is one of the smallest mammals in the world.

Origins and Relatives

Shrews, moles and voles are small, furry animals with short legs and tails. Their eyes and ears are rather small, and they have long whiskers. Some live in fields and gardens, close to humans. Others live in **savannas, prairies**, mountains and forests. A few even survive in hot deserts and the cold **tundra**.

Moles and shrews feed mainly on worms, insects and other small animals. Moles spend most of their lives in their underground tunnels. Shrews hunt mainly on the ground, hiding in burrows when they are not hunting. Voles look somewhat like shrews and moles, but they are not related to them. Voles are **rodents** that eat plants. Some feed on grass, while others eat seeds, fruits, shoots and even underground roots, bulbs and **tubers**.

Shrews, moles and voles belong to a large group of animals called **mammals**. They have soft, furry bodies, which are always warm to the touch, even in cold weather. Their young feed on milk produced by the mother.

The ancestors of shrews, moles and voles lived over 50 million years ago. **Fossils** of shrews 54 million years old have been found in North America. Today, there are more than 400 kinds of shrews, moles and voles.

Moles and shrews are closely related to each other. They have sharp, pointed teeth for feeding on insects and other small animals. Voles are related to rats, mice and squirrels. They have long, sharp front teeth for biting off grass and breaking open nuts and seeds.

The starnose mole uses its huge forefeet to tunnel through the soil in search of worms.

What Shrews Look Like

Shrews look similar to mice with narrow, pointed snouts and very long whiskers. Their eyes are tiny, and their ears are hidden in their fur. Their fur is usually brown or grayish, darker on the back than on the belly. They are very shy animals, darting into the undergrowth or down a burrow at the slightest sound.

The largest shrews, the African forest shrews, may grow up to 29 cm

An elephant shrew looks for ants among desert cacti, using its long, flexible snout.

(11 in) from head to tail. Most shrews are much smaller than this. The pygmy white-toothed shrew is the smallest mammal in the world. It is 3.5 cm (1.4 in) long, and weighs just 2 g (.07 oz). The European pygmy shrew is so small that it can use tunnels made by large beetles.

Shrews feed mainly on insects, worms and other small animals. Their teeth are very different from those of voles. They are sharp and pointed, and there is no gap between the **incisors** and the **molars**. The front teeth, or incisors, act like pincers to seize their **prey**. The molars are large and strong for crushing the shells of insects and the bones of frogs and small lizards.

Elephant shrews have long, flexible snouts, which they use for probing into cracks and loose soil for ants and **termites**. Elephant shrews have very long hind legs, and can hop and jump like little kangaroos.

Shrews are extremely nervous animals. Sudden loud sounds can literally frighten them to death. When frightened, their hearts may beat up to 1,200 times a minute.

Shrews have sharp, pointed teeth for crushing the bodies of insects and other small animals.

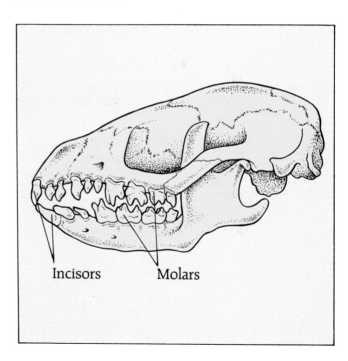

Incisors   Molars

What Moles Look Like

Moles are like fat bags of dark, velvety fur, with short legs and short tails. They have huge front paws, armed with long claws for digging. These paws are permanently turned outward, an ideal position for pushing

A hairy-tailed mole has very poor eyesight, but uses its sensitive snout to sniff and feel for food.

back soil as they tunnel. Their eyes and ears are so small that they are almost invisible, and their snouts are hairless and pink, with long whiskers.

Moles spend almost their entire lives underground. They dig out a series of tunnels, in which they live and hunt. Moles feed on worms, insects and other small animals that fall into their tunnels, so their diet is very similar to the diet of shrews. Their teeth are similar, too, but the

front teeth are not so long. Instead, the **canines**, the teeth just behind the front teeth, are long and sharp and overlap to help the mole seize and hold its struggling prey.

Sight is not much use underground. Moles are almost blind, and rely heavily on touch and smell to find their way around.

One group of moles, the desmans of southern Europe and parts of the USSR, has taken to hunting under water. A desman has waterproof fur, webbed feet and a long, flattened tail, which is used as a rudder. Its hind feet are larger than its front feet, and are used for paddling. It pokes its long, flexible snout above the water like a snorkel to breathe as it swims.

The desman uses its flexible snout as a snorkel when swimming.

What Voles Look Like

Voles look like large, fat mice with blunt noses. Their bodies are usually 10–11 cm (4 in) long, and their tails are quite short. They have small,

bright eyes, very small ears and long whiskers. Voles usually walk on all fours, but when they are feeding they often sit up on their hind legs and use their front paws to pick up and hold leaves, nuts and seeds.

Like rats and mice, voles have a pair of long sharp front teeth in each jaw. These teeth are called incisors. They are used like chisels to bite off grasses and scrape at nuts and seeds. The cheek teeth, or molars, are large and flat, with ridges of hard enamel for grinding up the food. Between the incisors and the molars is a wide gap. When it gnaws at its food, the vole draws its lips into this gap to prevent sharp pieces of broken-off food from going down its throat by mistake.

Most voles are gray, brown or reddish-brown in color. These

Left *This water vole's thick waterproof fur is still wet after its swim.*

colors blend with their surroundings, and make it difficult for their enemies to see them. The varying lemming, a vole that lives in the cold tundra, has a brown coat in summer, but molts to a white coat in winter to blend with the snow.

A short-tailed field vole nibbles at a grass blade with its long, sharp teeth. Voles also drink dew drops.

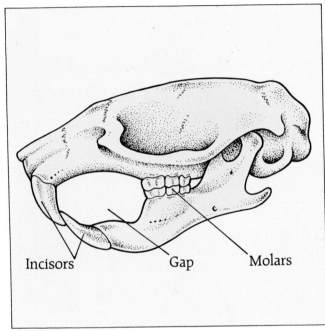

Incisors   Gap   Molars

A diagram of a vole skull. Voles have sharp incisors for chiseling nuts and seeds. The molars are used to grind food.

The fur contains two kinds of hairs. The short, soft hairs keep the animal warm, and the long, rough, guard hairs on the outside help to make the fur waterproof, and protect the skin.

2
Food and Feeding

A common shrew has found a grub among the moss.

Finding Food

Shrews, moles and voles rely mainly on touch and smell to find their food. Most moles hunt in the dark in underground tunnels. The whiskers on their faces and hairs on their feet and tails are very sensitive to touch. At the tip of the mole's bare snout are lots of little bumps, which are also sensitive to touch. When a prey animal falls into its tunnel, the mole detects the vibrations produced by its movements. It kills its prey by biting it. Some moles will attack quite large prey, such as mice and voles, using their claws as weapons.

Shrews hunt above ground, using their sharp teeth to seize and kill their prey. Water shrews hunt under water, but they drag their prey out of the water to kill it. The desman also hunts under water, using its snout to find leeches, snails and worms in the mud.

Some shrews have a poisonous bite that paralyzes or kills their prey. This enables them to catch quite large animals. The short-tailed shrew can overpower a large mouse. Shrews are very fierce little animals, ready to fight intruders. They usually drag their prey into a burrow or other hiding place before eating it.

The bank vole holds a blackberry in its paws as it feeds.

The starnose mole has 22 pink finger-like feelers at the top of its snout. These feelers are very sensitive to touch, and the mole waves them as it hunts.

Most voles feed above ground, grazing on grass, or eating flowers, leaves, seeds, nuts and berries. A few voles, such as the lemmings, make tunnels underground to dig for roots, **bulbs, corms** and tubers.

How Shrews, Moles and Voles Feed

Shrews, moles and voles are very active animals and use a lot of energy. They burn energy very rapidly and act like small eating machines. They can eat their own weight of food every day. This would be the same as a 23-kg (50-lbs) child eating 200 hamburgers a day.

Some shrews feed almost all the time. They digest their food very rapidly. Shrews and voles will sometimes eat their own droppings (undigested food) to make sure they get all possible nutrients out of their food. The insects they eat have very hard shells, and this causes a lot of wear and tear on the teeth. Many adult shrews die of starvation because their teeth have worn down and they can no longer feed properly.

Voles' teeth also wear down, but their teeth keep growing throughout the vole's life, so they never become too worn down. A vole's gut contains millions of **bacteria**, which help to break down tough plant material so the vole can digest it.

Moles, voles and shrews will sometimes store food when it is scarce, or when winter is coming. One

This shrew is cleaning its anus, eating undigested food that it has just passed out of its body. The food will pass through its gut again, so more of it can be digested.

mole's food store was found to contain 18 grubs and 1,280 earthworms, neatly arranged in piles of about ten worms. The mole had bitten off the worms' heads to make sure they did not get away. Voles store seeds, nuts and grass in their burrows or in cracks in rocks, or hide

When food supplies are scarce, or winter approaches, the bank vole stores nuts in its burrow.

them under piles of leaves. Mountain voles make haystacks and build little walls of stones around them to keep the hay from blowing away.

3
Where Shrews, Moles and Voles Live

The field vole is found in grassy places throughout Europe, the USSR and Central Asia.

Grasslands and Mountains

Since many voles feed mainly on grass, they can usually be found in grassy places – prairies, savannas, **steppes**, meadows and fields. Many wild flowers growing in grasslands have juicy underground parts, such as roots, bulbs, corms and tubers. The voles dig for these or make underground tunnels to reach them.

Voles also make burrows in which to sleep and hide from their enemies. A few feet below the ground the temperature does not vary much, so a burrow provides good protection against the heat of summer and the cold of winter.

Moles and shrews also like grasslands, as there are plenty of worms and insects there for them to eat. Shrews prefer damp places, but moles will tunnel wherever the soil is not too dry and hard or wet and soggy. You can see where a mole lives in grassland by the molehills it makes as it tunnels.

Shrews and voles may also live quite high up on mountains, provided there is plenty of shelter in the form of bushes or rock crevices. The field voles can be found up to 1,800 m (6,000 ft) above sea level in alpine meadows. The snow vole of Europe and Asia lives even higher, up to 4,000 m (13,000 ft). Where the soil is too wet to make a burrow, it makes its home in cracks in the rocks.

The California mole tunnels just below the surface in an alpine meadow, throwing up a long ridge of soil behind it as it goes.

Tundra and Deserts

Some voles and shrews live on the vast, treeless plains of the far north. The lemming, a kind of vole, feeds on the underground parts of plants. It has long claws on its front feet for digging. When the ground is frozen and snow covered in winter, it makes tunnels in the snow, nibbling at any plants it finds. The snow protects it from the cold air above, and from the eyes of its enemies.

The collared lemming develops horny shields on the third and fourth claws of its forefeet in winter. These help it to dig in the frozen snow, and protect against frostbite. It is sometimes called the "hoofed lemming."

Lemmings have very loose, saggy skin. When they are cold, they fluff up their fur and snuggle down into their own skin. Fur around their eyes,

The common shrew stays active all year round, even in cold weather.

nostrils and ears helps to keep the snow out.

Because the food supply in deserts is unreliable, very few shrews, moles and voles live in them. Plants grow only after the rare showers of rain, and without plants, there are few insects.

Some golden moles live in sandy deserts, burrowing just under the loose sand, leaving long ridges behind them. The piebald shrews of the European steppes have fringes of stiff hairs on their feet to help them run on loose sand. Voles called mole voles and mole lemmings also live in dry places. These are burrowing voles with velvety mole-like coats.

Shrews are fierce animals, and will even attack animals larger than themselves. This American desert shrew has caught a lizard.

Forest and Woodlands, Rivers and Ponds

Woodlands and forests have plenty of food in the form of insect and plant life. Moles may live in woodlands where the soil is moist and easy to

A pygmy shrew sleeps curled up on a bed of moss.

burrow in. Shrews and voles find plenty of shelter here, under leaves, among tree roots, and in the "borrowed" burrows of other animals.

The pine vole of northern forests is so unwilling to be away from its shelter that it takes all its food back to its burrow to eat. It even covers its runways with fallen leaves and grasses to hide them.

The heather voles of Canada and the United States are good at climbing trees. The males live in burrows in the ground, but the females build nests in the trees.

Some voles, moles and shrews are very good swimmers. Desmans are found in ponds and slow-flowing rivers in southern Europe and the USSR. They live in burrows in the river bank with entrances below the water level. This protects them from land enemies such as foxes and weasels. Water voles and water

shrews also have underwater entrances to their burrows. They sometimes use old mole burrows and dig their own tunnels to the water.

The water shrew does most of its hunting in the water, feeding on water snails, mud-dwelling worms, insect **larvae** and other small water animals.

A northern water shrew emerges from the water. It will rub its fur against the earth walls of its burrow to dry itself.

The water voles use the water only to get from one good patch of plants to another, and to escape from their enemies.

4
On the Move

Air trapped under the water vole's fur helps to keep it afloat as it swims.

Walking, Running and Swimming

Despite their short legs, voles and shrews are quite agile. They need to be able to run fast to escape enemies such as owls, hawks and foxes. Many voles and shrews can climb bushes, and even trees, using their claws to grip the bark. They can also jump. European meadow voles can leap a 20-cm (8-in) wall to escape danger. The elephant shrews, which have long hind legs, often hop along like kangaroos. Moles do not move very fast, as their paddle-shaped front paws cannot lift their bodies off the ground very well.

Many voles and shrews are good swimmers, and so are the desmans. They use their hind feet to propel themselves through the water. Webs of skin between the toes, or fringes of stiff hairs, help to increase the area of the foot for pushing against the water.

animal looks silvery white as it dives in search of prey. Water shrews use their tunnels to dry their fur, rubbing themselves against the sides and letting the soil soak up the moisture.

The water shrew has fringes of stiff bristles along its toes to make its feet more like paddles.

A bank vole stands on its hind legs while it looks around for danger.

The tail may also be fringed with hairs, and is used as a rudder. The desman can close its nostrils and ears with little flaps to keep the water out.

All these animals have waterproof fur. The lemming's fur is so thick that it traps a lot of air. When the lemming swims, the air helps it to float, so its back stays above the water. The water shrew's coat also traps air, and the

Burrowing

Most shrews, moles and voles make some sort of burrow, using their claws for digging. They have small ears, which do not get in the way in narrow tunnels, and their fur is soft, so it does not get too ruffled as it rubs against the wall of the burrow. Voles and shrews sometimes use the abandoned burrows of other animals.

Moles are built for burrowing, with their large, front paws. Although their bodies are heavy and flabby, this does not matter, since they are usually supported by the floor of the tunnel. Moles have soft, velvety fur, which can be smoothed down to lie flat either forward or backward. This helps the mole to move to and fro in its tunnel without ruffling its fur.

When tunneling, the mole draws in its head and uses its front paws one after the other to tear open the ground, almost as if it is swimming through the soil. It throws the soil behind it, using its body to press the soil firmly against the sides of the tunnel. From time to time it digs a tunnel to the surface and pushes out the excess soil to form a molehill.

A European mole digging. The mole's feet face sideways to push the soil aside as it burrows. Its velvety fur will lie flat either forward or backward, so it can turn around in its tunnel without ruffling its coat.

Burrowing voles usually have very long claws on their front paws for digging. The collared lemming grows extra long claws on the third and fourth fingers of its front paws in winter, to help it dig in the frozen snow. The mole lemmings and mole voles use their large front teeth instead of their claws for digging.

Moles have strong, spade-like feet armed with long claws for digging. The stiff bristles on the mole's snout help the animal to feel its way in its dark tunnels.

Wandering Lemmings

Some voles can travel long distances. From time to time, lemmings multiply too fast, and, because there are too many of them, they start to run out of food. When they become too crowded, the lemmings get very excited and restless, and leave their area on the tundra in huge numbers. They travel away from the tundra toward the birch forests farther south, where there will be more food. The lemmings rush along in a large crowd as if they are in a panic, crossing mountains and rivers, not even stopping to feed on the way. Sometimes the leading lemmings may reach a cliff or the edge of a lake or the sea, where the force of the hurrying crowd behind pushes them over the edge. Large numbers of lemmings may fall to their deaths or drown. But the few who survive may find a new home where food is more plentiful.

Left *In winter, the Arctic lemming's coat turns grayish white to blend with the snow and frost. This helps to hide the lemming from animals that hunt them.*

Right *Collared lemmings live on the Arctic tundra of northern Europe, Siberia and North America. They feed on low-growing plants, roots and tubers.*

5
Family Life

A common shrew sniffs at a log to find out what other shrews have visited it.

Social Life

Most moles are very unsociable animals, and prefer to live alone. If several moles are put in a cage, they will fight and sometimes kill each other. However, when there is a plentiful supply of food, European moles may share a large tunnel system. Desmans often share dens.

Shrews are also unsociable and usually live alone. A shrew will defend the piece of land in which it lives and feeds. This is called its territory. It will attack any shrew entering its territory, rising up on its hind legs and lashing out with teeth and claws. When a shrew thinks it is losing a fight, it will lie on its back and scream. The other shrew will stop fighting and allow it to run away. By defending its territory, the shrew makes sure that it will be able to find enough food in the area around its burrow.

Shrews have extremely good hearing, and can hear sounds that are too **high-pitched** for humans to hear. They make squeaks and high-pitched clicking sounds that can tell other shrews how strong they are, and may help to settle disputes without the need for a fight. Voles are more sociable than shrews and moles, and often share tunnel systems.

These wandering shrews are having a fierce battle over the rights to a piece of territory.

A pygmy shrew rears up on its hind legs and raises its nose to sniff the air. It can detect the presence of other shrews, or of enemies, by their smell.

Moles, voles and shrews use smells as well as sounds to "talk" to each other. They mark out their territories by smearing a smelly liquid around its borders. This tells other animals that someone else lives there, and prevents a lot of unnecessary fighting.

Underground Homes

Shrews and voles have many enemies, so they need sheltered homes hidden from view. Some make nests of dried leaves and grasses in hollows under tree roots or in cracks in rocks, but most use underground burrows. Sometimes they use the old burrows of other animals, but usually they dig their own. Water voles and desmans often make their burrows in the banks of streams and lakes, with the entrances under water. A few voles make their nests in bushes or trees.

Moles hunt, eat and sleep underground. They dig complex tunnel systems. The European mole

It is easy to see water voles along the banks of rivers and streams near their burrows.

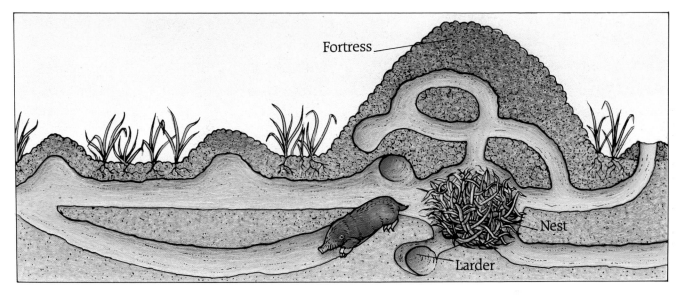

Fortress

Nest

Larder

The European mole digs out an elaborate series of underground tunnels, in which it hunts for worms.

usually makes two almost circular tunnels, one inside the other, linked together by another tunnel. Side tunnels lead off the main circles to the mole's main hunting grounds. One circle is deeper than the other, and grass-lined rooms leading off are used for resting and as a nursery. In wet soil, the mole may make an extra large molehill and make its nesting den inside, to keep dry.

Voles and shrews make regular runways from their burrows to their favorite feeding areas. By always using the same route, they flatten the grass into a tiny path. This also helps the animal escape if danger threatens – it is much quicker to flee along a path than it is to push through dense tangles of grass.

Courtship and Mating

For a solitary animal like the mole, meeting females is difficult. Special behavior, called **courtship**, is needed to overcome the two animals' suspicion of each other. In spring, the mole leaves his burrow at night and goes visiting. When he enters a female mole's burrow, the two animals may fight at first, but eventually they get used to each other, and then they **mate**.

Shrews overcome each other's desire to fight by giving off special male and female smells. When the female shrew wants a mate, she keeps

A pair of voles mating. After mating, the male will leave the female.

calling "peep, peep, peep." The male is attracted by her calls and by her smell. At first he keeps advancing and retreating, unsure of her reaction. He may even try to fight with her, but eventually he begins sniffing her and making excited twittering sounds. The female starts to walk around holding her rump in the air, inviting him to mate with her.

Courting water shrews chase each other through the water before they mate, while male common shrews may drag their females into the bushes to mate. Voles also use sounds and smells to attract each other.

Most moles, voles and shrews make very bad fathers. The male leaves the female after mating, and wanders off to find other females. The female gives birth and brings up the young quite alone. Only the star-nosed mole and a few kinds of voles live in pairs and bring up the family together.

Right *This pair of adult bank voles have met among the leaves on the woodland floor. Soon they will mate. The female will bring up the family on her own.*

Bringing up the Family

Voles and shrews can multiply extremely rapidly, producing up to ten young at a time, sometimes ten times a year. One vole is recorded as having

These young moles are cozy and warm in their nest deep under ground inside the nursery in their mother's burrow.

Common shrews are born naked and blind, but they can still find their mother and suck milk from her nipples.

produced 127 young during her short life. The young themselves are often ready to breed when they are only a few weeks old. Moles usually produce only a few young once a year, but they may live for several years. Voles and shrews often live for only a year or two.

Female moles, voles and shrews make special nursery nests of soft grass and leaves in their burrows. Here, they give birth to their babies. The young are born completely helpless: blind, deaf, naked and unable to walk. Newborn shrews weigh only 1 g (0.04 oz). When they need their mother they make tiny cries. A mother can recognize her own young by their smell.

A line of shrews.

The young feed by sucking milk from nipples on their mother's belly. They grow fast, and soon develop furry coats. They start to crawl around the nursery, feeling their way with their long whiskers. It may be one or two weeks before their eyes open. Eventually they are big enough to venture outside with their mother. They still feed on her milk until they have learned how to find food for themselves.

If danger threatens, shrews and voles may pick up their young in their mouths and carry them to a safe place. Young shrews form a long line behind their mother, each baby clinging to the fur of the one in front.

6
Enemies of Shrews, Moles and Voles

This barn owl is taking a vole home to feed its young.

Shrews, moles and voles have many enemies. Foxes, cats and weasels hunt them by day. Hawks swoop on them from the air, and snakes attack them in their own burrows. At night, owls hunt them. Even in the water they are not safe. A water shrew makes a tasty meal for a heron or a large fish like a pike.

Their best defense is their color, which blends with their surroundings, making them difficult to see. They are also extremely cautious, darting under a bush or stone at the slightest sound. Because they usually stay in their own home territory, they are seldom far from a burrow entrance.

Many moles and shrews give off an unpleasant smell, and few animals will eat them, even if they are able to catch them. Many animals avoid these shrews and moles altogether because they are not good to eat.

Farmers dislike moles because they make molehills among the crops. These moles have all been shot.

The armored shrew from tropical Africa is incredibly strong. It has a special tough backbone, and can survive being stepped on by a full-grown man. Some Africans believe it is magical. They think that if you eat an armored shrew, you will become brave and cannot be harmed. Some golden moles lie down and pretend to be dead if danger threatens.

Shrews, moles and voles use their good hearing and powerful sense of smell to detect their enemies in time to make their escape. They will often utter warning cries to each other when danger threatens. Brandt's voles stand up on their hind legs and whistle a warning.

A fox watches a shrew in the grass, waiting for the right moment to pounce.

7
Learning More About Shrews, Moles and Voles

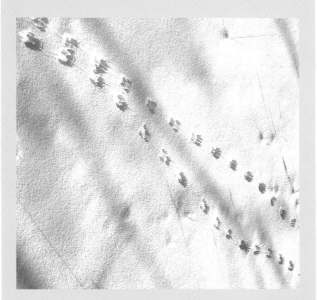

Voles are very shy animals and difficult to see, but you can often find their tracks in snow or in damp mud.

Voles and shrews are very difficult to see because they are so shy. They can hear or smell you coming and hide long before you see them. But it is easy to find out where they live. If you walk through grassy fields, you may find the entrances to a burrow. If a burrow is occupied, there may be droppings nearby. If you sit very still and quiet for a long time near a burrow, the animal may emerge.

If you search carefully, you may well see tiny paths of flattened grasses leading from the burrow to favorite feeding areas. Voles cut the grass short as they feed, so you may find tiny patches of short grass.

In Europe, water voles are some of the easiest voles to see. As you approach them along a river bank, they drop into the water with a loud "plop," and you can watch them swimming away. You may find them sitting on the bank nibbling grass.

Voles will sometimes strip bark from trees and shrubs. Look for the scrape marks of teeth, which show up low down on the trunk.

Molehills are a common sight on hills and in fields and meadows. Old molehills may be covered with grass, but new ones will be of fresh, brown

If you hear a sudden "plop" as you walk along the river bank, look for a water vole swimming away.

Molehills are a common sight in damp meadows. Sometimes you can trace the pattern of a burrow by following the line of molehills.

earth. You can try to guess the pattern of the tunnels below. A single mole may produce up to a hundred molehills. Can you spot the extra large hill where the mole has its nest?

Glossary

Bacteria Minute, living creatures too small to see.

Bulb A swollen underground bud-like plant structure, composed of fleshy leaves in which food is stored.

Canines Large, pointed teeth just behind the front teeth.

Corm A short, swollen underground stem in which food is stored.

Courtship The way in which male and female animals behave before mating.

Fossils The remains of animals and plants that died long ago and are now preserved in stone.

High-pitched Sounding very high and squeaky.

Incisors The teeth at the front of the jaw.

Larva (plural, larvae) A young animal that is different in shape from its parents.

Mammals Warm-blooded animals with hair or fur, the females of which produce milk to feed their young.

Mating Joining together as male and female to produce young.

Molars Large teeth at the back of the jaw.

Prairies Grasslands found in North America.

Prey An animal that is killed and eaten as food by another animal.

Rodents A group of gnawing or nibbling mammals that includes rats, mice and squirrels.

Savannas Grasslands found in Africa.

Steppes Grasslands found a long way from the sea in Europe and Asia.

Termites Large ant-like insects found in warm climates.

Tubers Underground stems very swollen with stored food.

Tundra Large treeless plains found in the far north.

Finding Out More

If you would like to find out more about shrews, moles and voles, you can read the following books:

Grahame, Kenneth. *The Wind in the Willows*, Scholastic, 1988

Whitaker, John O. *The Audubon Society Field Guide to North American Mammals*. Alfred A. Knopf, 1980

Wild Animals of North America. The National Geographic Society, 1979

Picture acknowledgments

All photographs from Oxford Scientific Films by the following photographers: Caroline Aitzetmuller 28; G. Bernard 15; Attilio Calegari 22; Jack Dermid 41 (right); Michael Dick (Animals Animals) 10; Mark Hamblin 40; Rodger Jackman 19; Breck P. Kent (Animals Animals) 9; Michael Leach 30, 43; Zig Leszczynski (Animals Animals) 12, 17 (right); Ted Levin (Animals Animals) 31; G.A. Maclean 34; Mendez (Animals Animals) 23; Press-tige Pictures 14, 16, 20, 26, 27, 29, 37; Wilf Schurig (Animals Animals) 42; Tim Shepherd 33 (right), 43 (right); Stouffer Enterprises (Animals Animals) 25, 27 (right), 33 (left); David Thompson 38 (left), 41 (left); Tony Tilford 8; Barrie E. Watts 17 (left), 18, 24, 32; Jack Wilburn (Animals Animals) 21.

Index

Pages for illustrations are shown in **bold** type.

DATE DUE

OCT 26 '90			